Complete the number sentences.
Draw a ring round the numeral.

1

3 + 2 = →   1   2   3   4   (5)   6   7   8   9

6 + 1 = →   1   2   3   4   5   6   7   8   9

4 + 5 = →   1   2   3   4   5   6   7   8   9

1 + 6 = →   1   2   3   4   5   6   7   8   9

2 + 1 = →   1   2   3   4   5   6   7   8   9

7 + 2 = →   1   2   3   4   5   6   7   8   9

6 + 2 = →   1   2   3   4   5   6   7   8   9

1 + 3 = →   1   2   3   4   5   6   7   8   9

2   Complete the number sentences.
    Draw a ring, round the numeral.

		1 2 3			1 2 3
3+5 =>	4 5 6		3+6 =>	4 5 6
	7 (8) 9			7 8 9

	1 2 3			1 2 3
4+2 =>	4 5 6		1+2 =>	4 5 6
	7 8 9			7 8 9

	1 2 3			1 2 3
5+1 =>	4 5 6		5+3 =>	4 5 6
	7 8 9			7 8 9

	1 2 3			1 2 3
1+5 =>	4 5 6		4+4 =>	4 5 6
	7 8 9			7 8 9

Complete the number sentences.
Draw a ring round the numeral.

5+2= ➤
```
    1   2   3
    4   5   6
    7   8   9
```

4+3= ➤
```
    1   2   3
    4   5   6
    7   8   9
```

1+8= ➤
```
    1   2   3
    4   5   6
    7   8   9
```

3+4= ➤
```
    1   2   3
    4   5   6
    7   8   9
```

7+1= ➤
```
    1   2   3
    4   5   6
    7   8   9
```

2+2= ➤
```
    1   2   3
    4   5   6
    7   8   9
```

5+4= ➤
```
    1   2   3
    4   5   6
    7   8   9
```

6+3= ➤
```
    1   2   3
    4   5   6
    7   8   9
```

4   Ring the sets. There are ten (10) members in each.

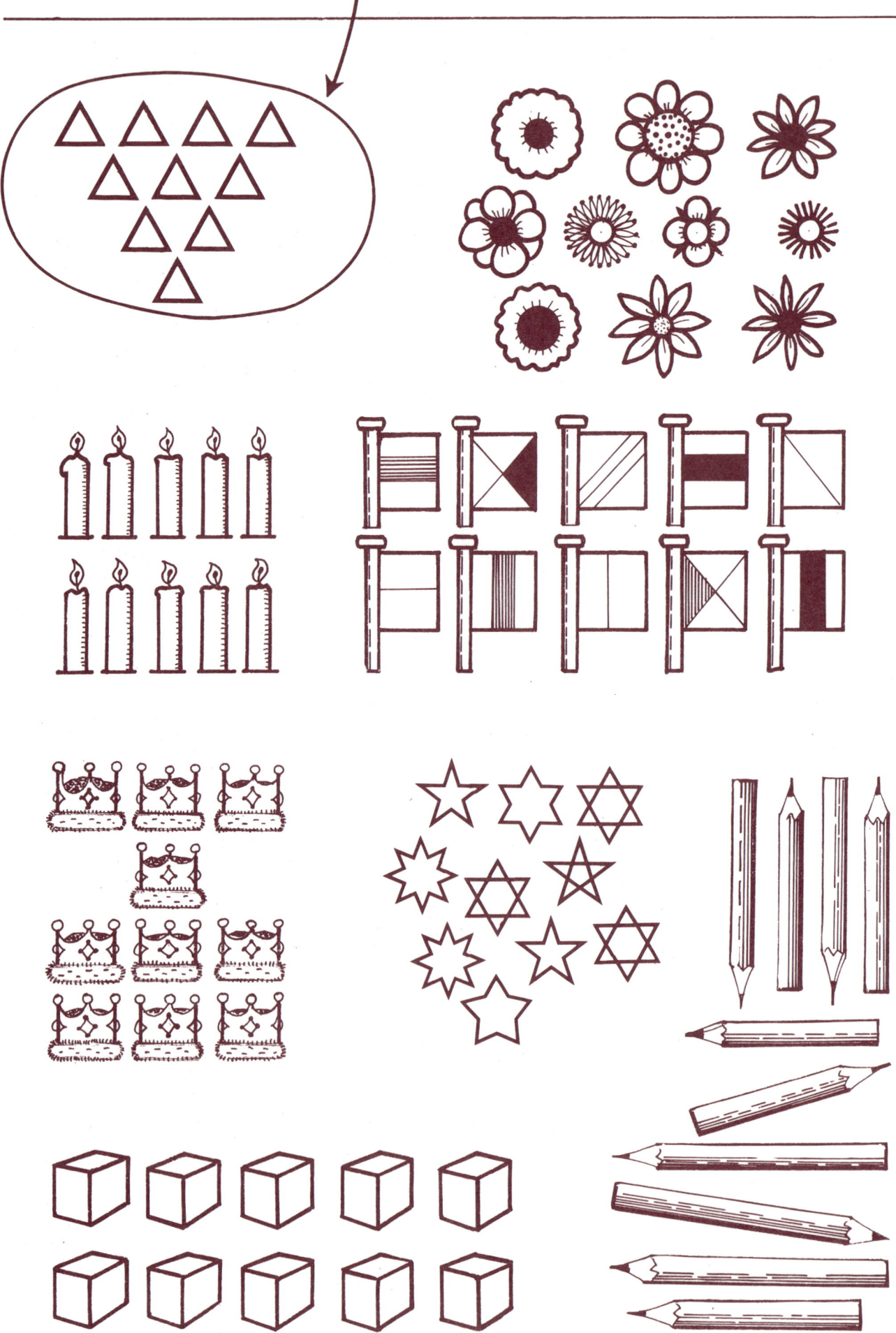

Ring the sets. There are ten (10) members in each.

5

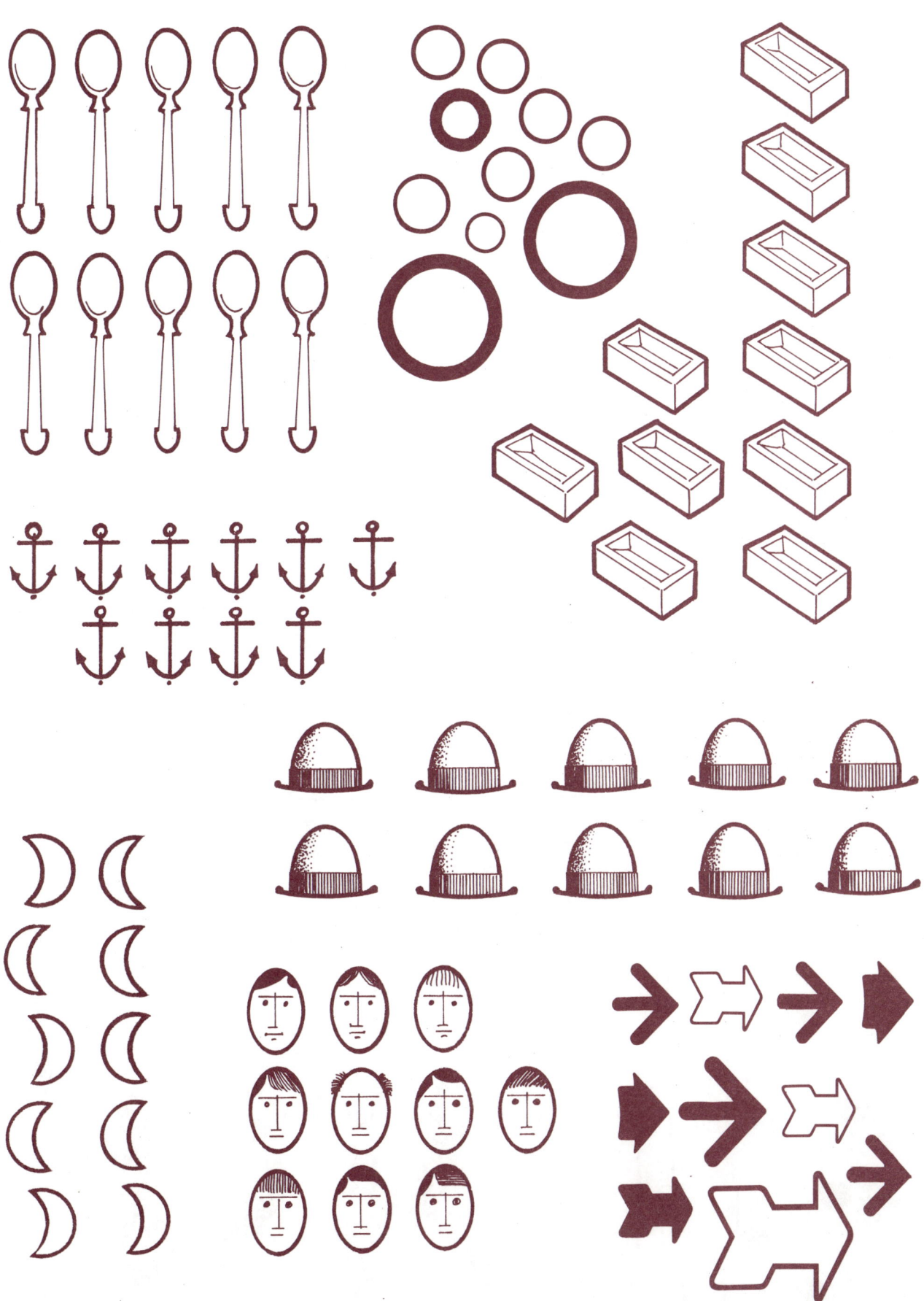

Colour the sets that have ten (10) members.

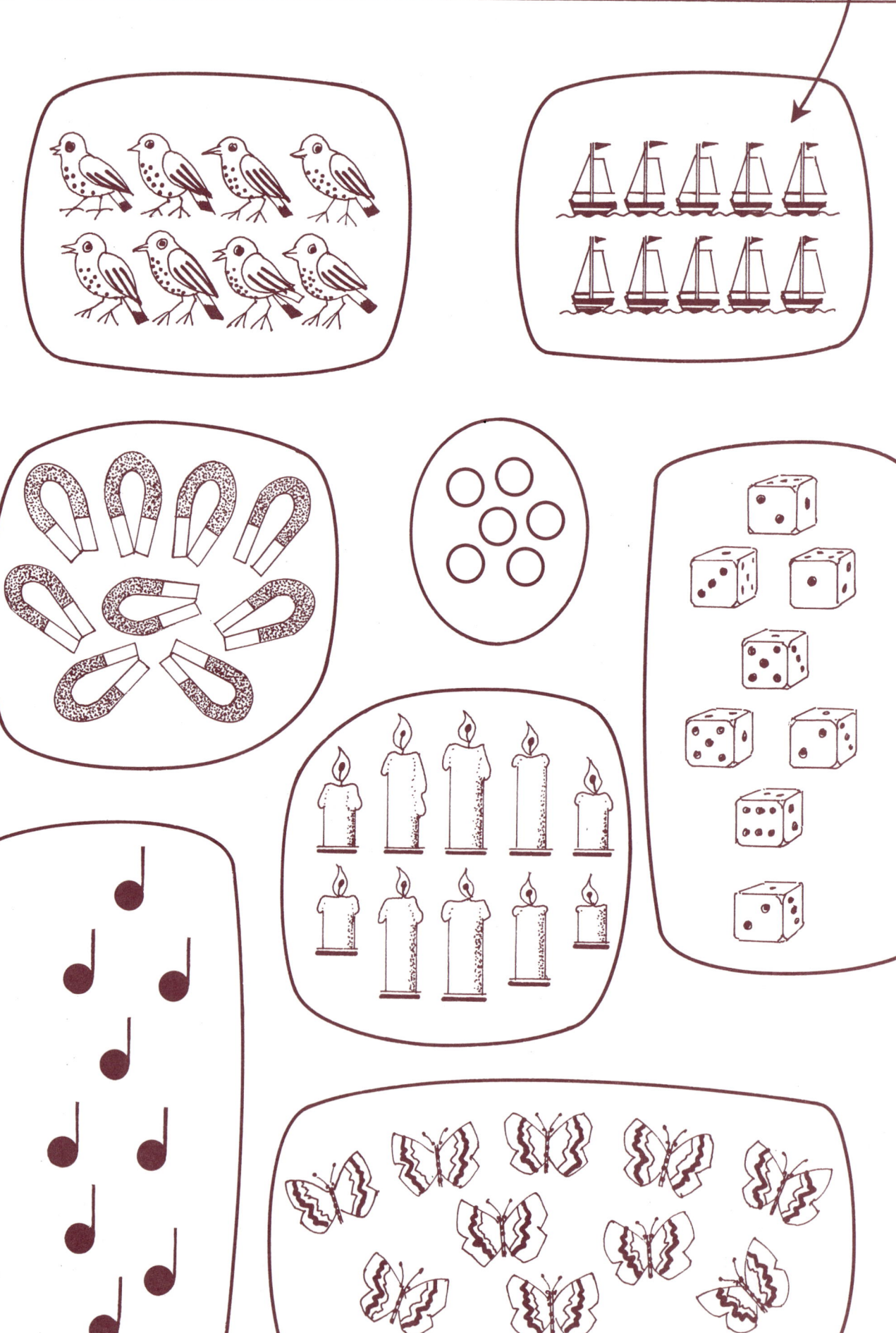

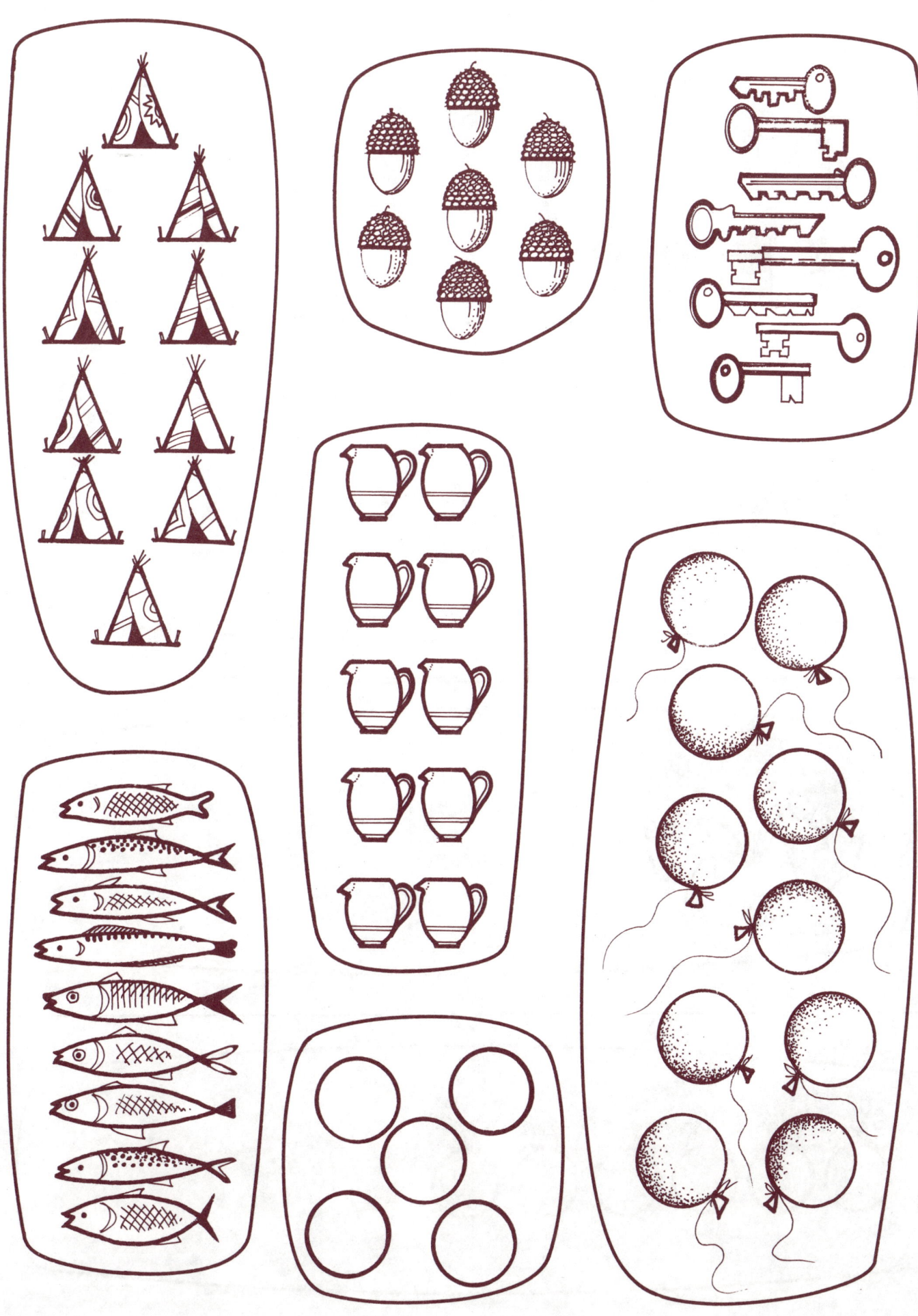

8   Make each set into a set of ten (10).

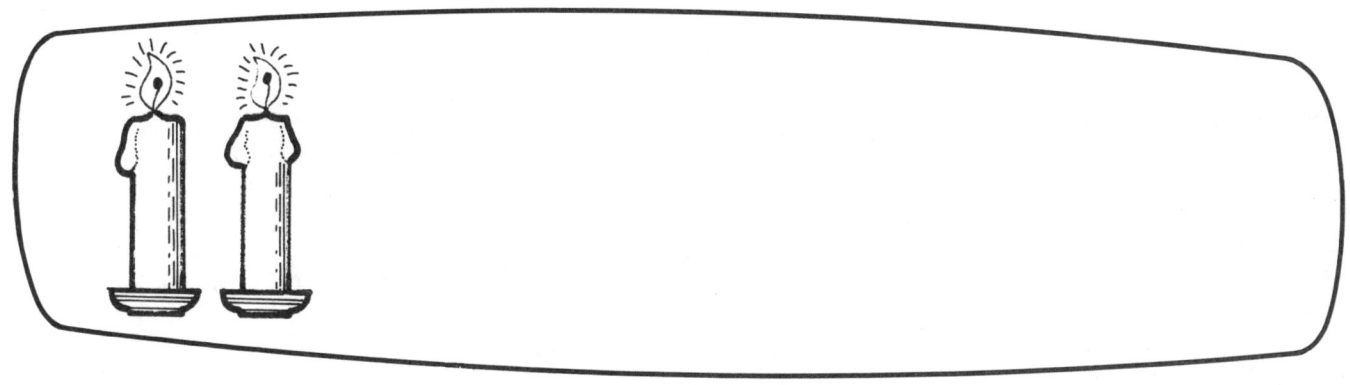

10 Colour the sets.

Write the numerals.   Write the words.

4

four

10

ten

Colour the sets.
Write the numerals. Write the words.

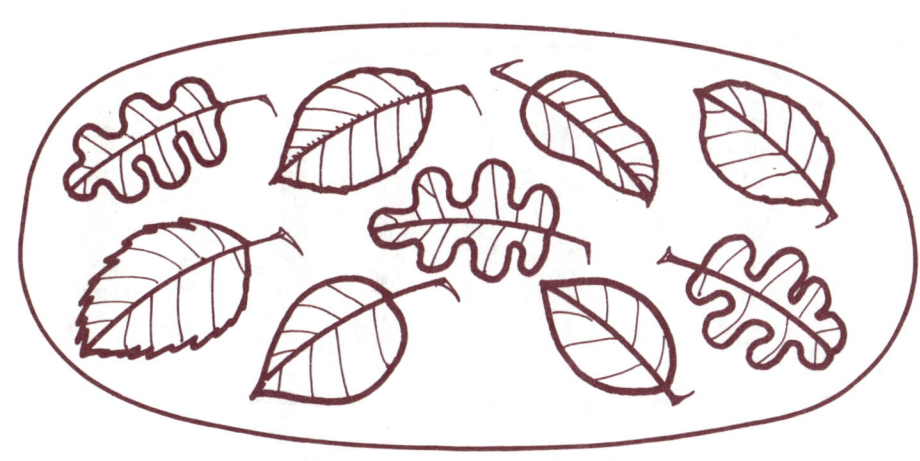

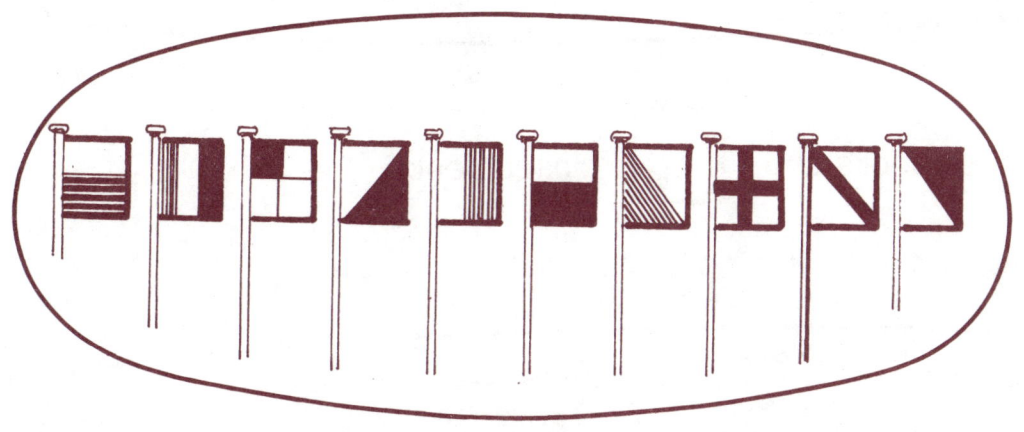

12 Partition the sets.
Write the numerals.

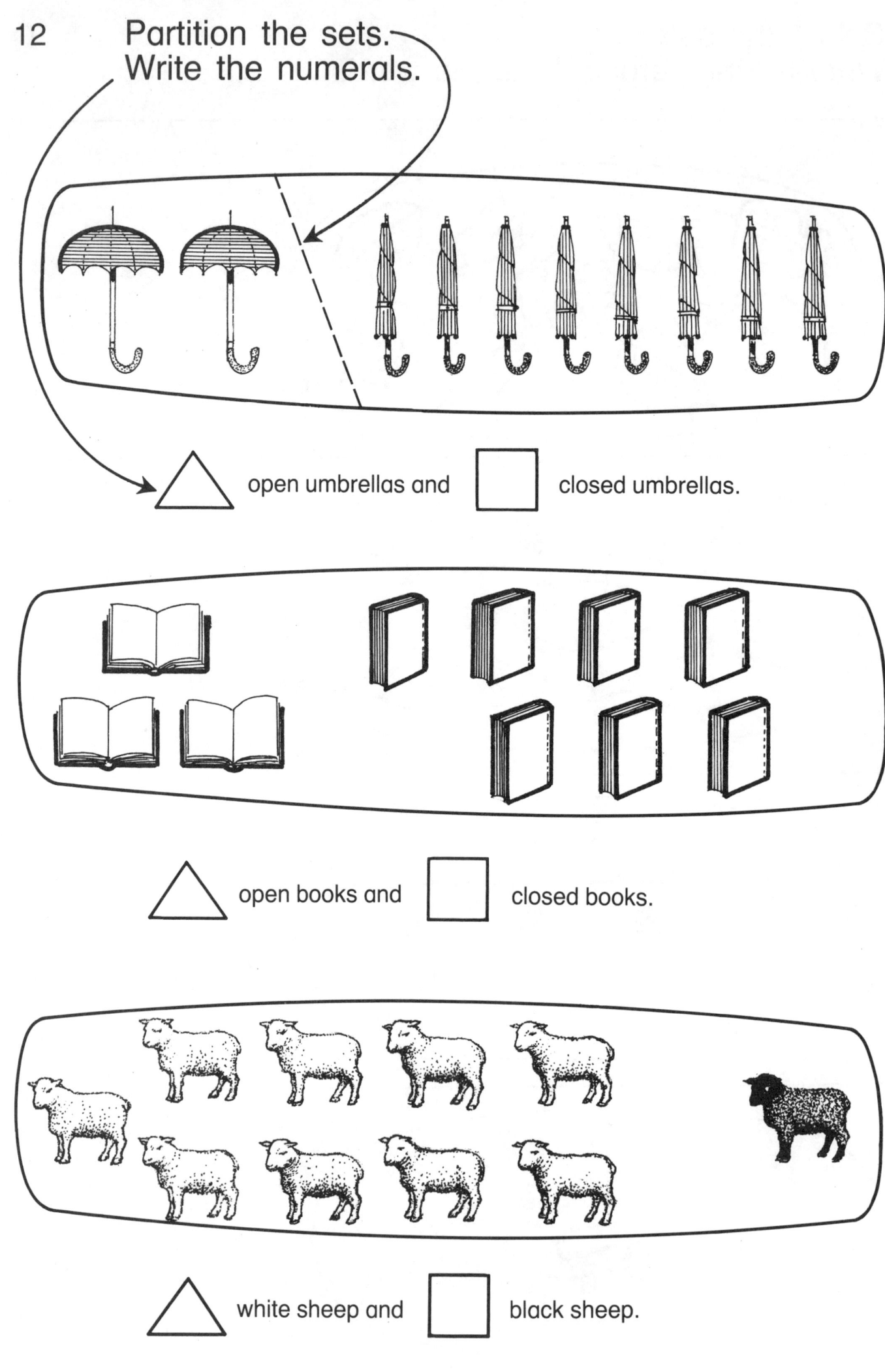

△ open umbrellas and ☐ closed umbrellas.

△ open books and ☐ closed books.

△ white sheep and ☐ black sheep.

Write the numerals.

There are  open umbrellas.

There are  closed umbrellas.

There are  umbrellas altogether.

 umbrellas plus  umbrellas equals  umbrellas.

$$2 + 8 = \langle\ \rangle$$

There are  open books.

There are  closed books.

There are  books altogether.

 books plus  books equals  books.

$$3 + 7 = \langle\ \rangle$$

There are  white sheep.

There is  black sheep.

There are  sheep altogether.

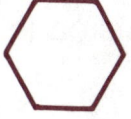 sheep plus sheep equals sheep.

$$9 + 1 = \langle\ \rangle$$

14   **Write the numerals.**

There are 8 oak trees in the field.

There are 2 ash trees in the field.

There are ⬡ trees altogether

8 trees plus 2 trees equals ⬡ trees.

$$8 + 2 = \hexagon$$

There is 1 oval plate on the table.

There are 9 round plates on the table.

There are ⬡ plates altogether.

1 plate plus 9 plates equals ⬡ plates.

$$1 + 9 = \hexagon$$

Diane has 5 sweets.

Adam has 5 sweets.

They have ⬡ sweets altogether.

5 sweets plus 5 sweets equals ⬡ sweets.

$$5 + 5 = \hexagon$$

Write the numerals.

There are 4 classrooms upstairs.

There are 6 classrooms downstairs.

There are ⬡ classrooms altogether.

4 classrooms plus 6 classrooms equals ⬡ classrooms.

$$4 + 6 = \hexagon$$

We won 7 of our matches.

We lost 3 of our matches.

We played ⬡ matches altogether.

7 matches plus 3 matches equals ⬡ matches.

$$7 + 3 = \hexagon$$

There are 6 plain cakes on the table.

There are 4 fancy cakes on the table.

There are ⬡ cakes altogether.

6 cakes plus 4 cakes equals ⬡ cakes.

$$6 + 4 = \hexagon$$

16 Colour the sets.
Write the numerals. Write the words.

7
seven

0
zero

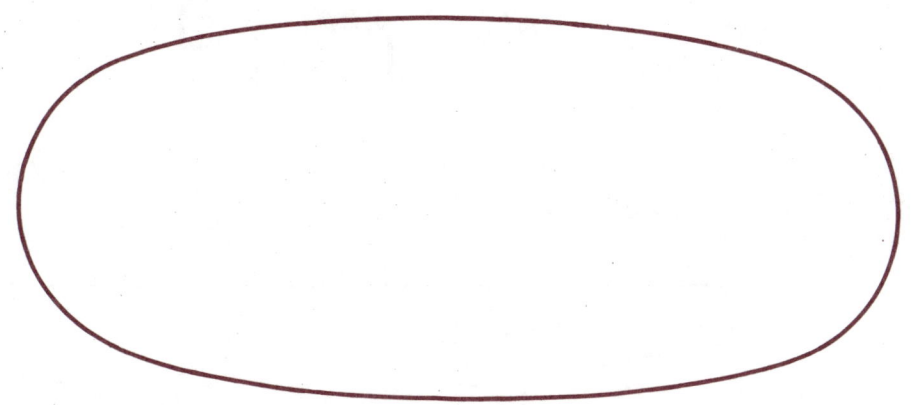

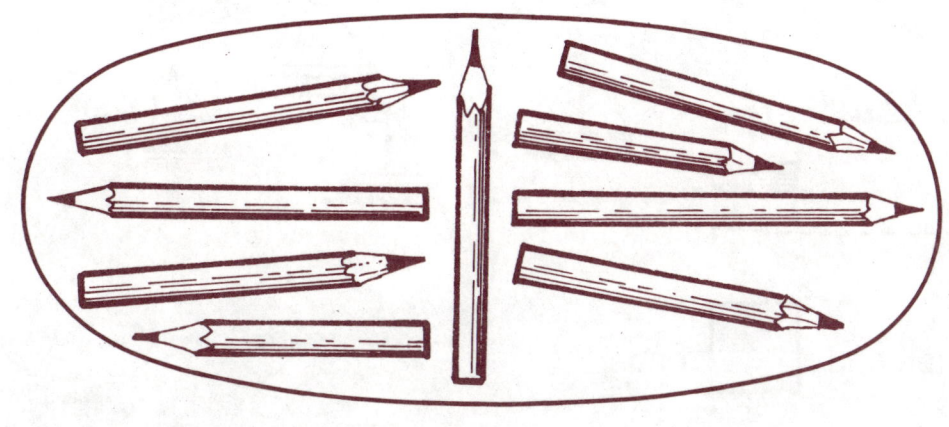

Ring a set of ten (10). Write the numeral.

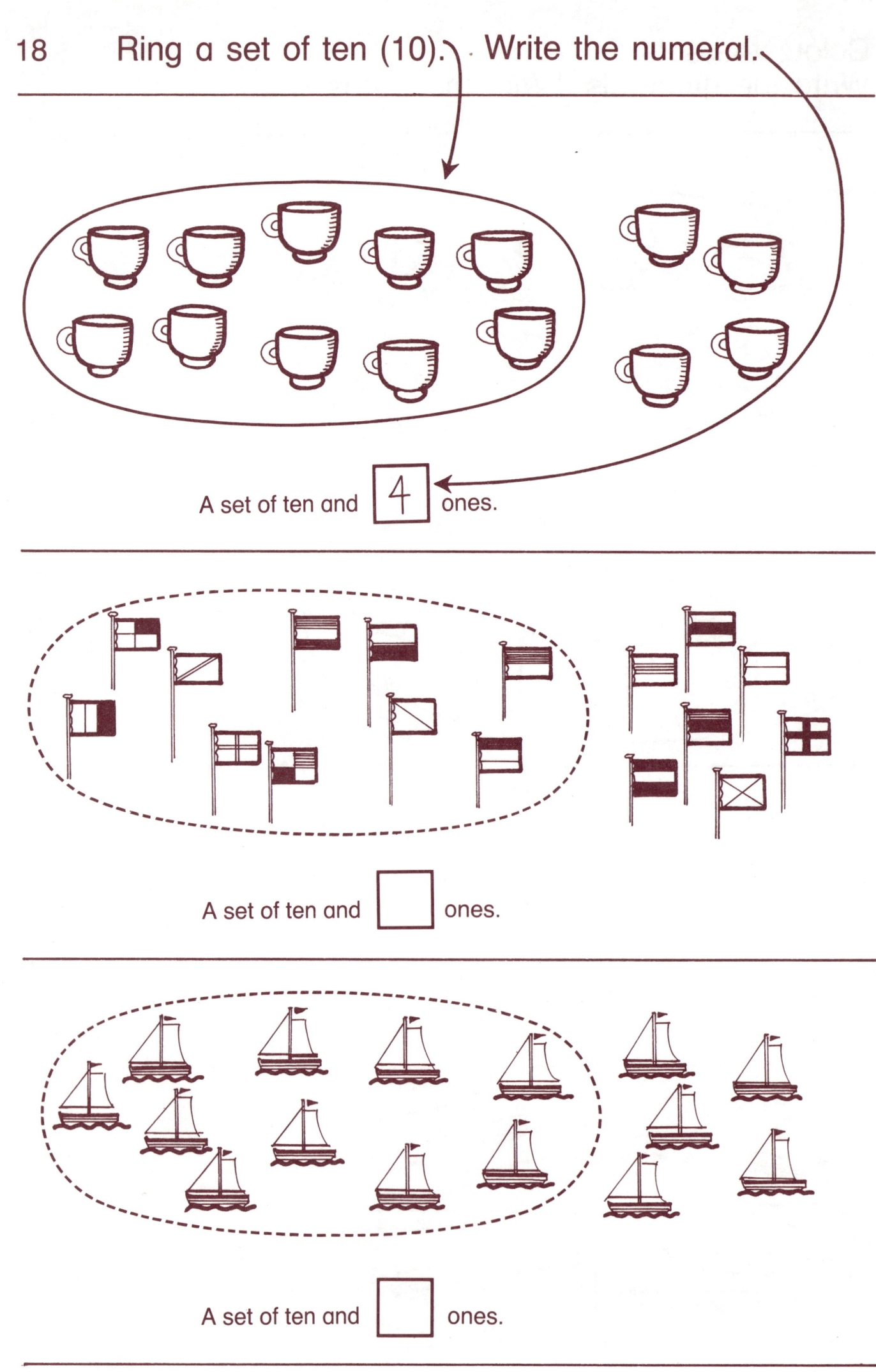

A set of ten and $\boxed{4}$ ones.

A set of ten and $\boxed{}$ ones.

A set of ten and $\boxed{}$ ones.

A set of ten and ☐ ones.

A set of ten and ☐ ones.

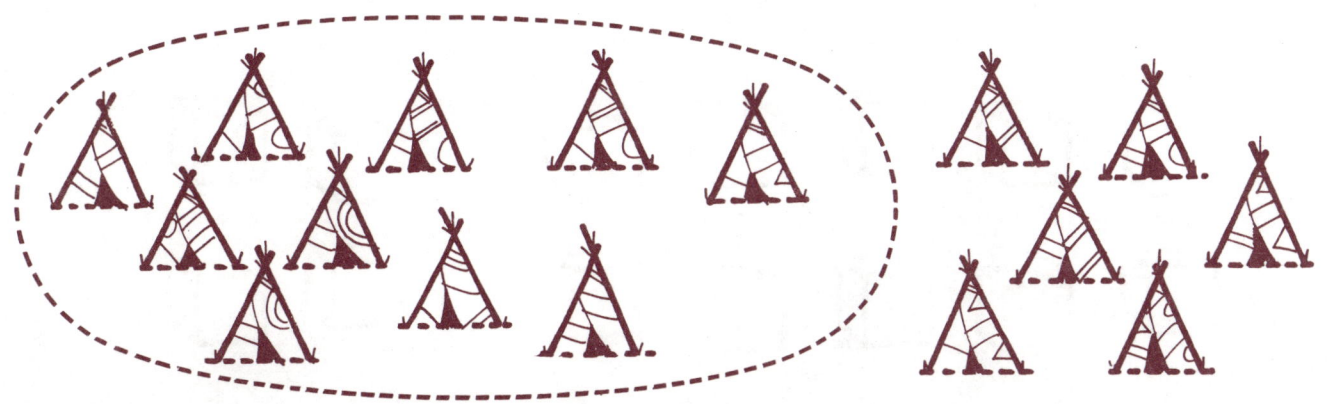

A set of ten and ☐ ones.

Ring a set of ten (10). Write the numeral.

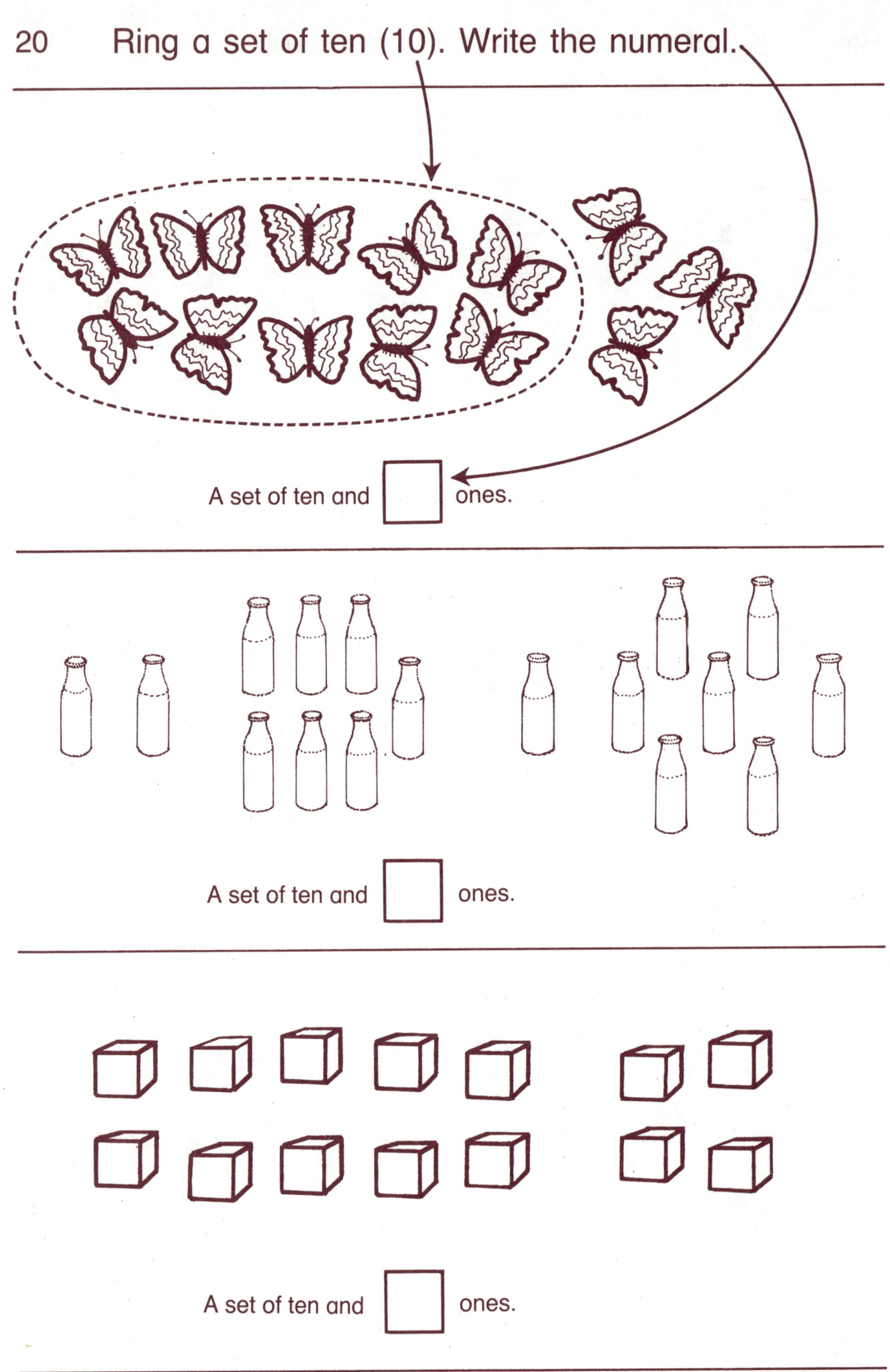

A set of ten and ☐ ones.

A set of ten and ☐ ones.

A set of ten and ☐ ones.

A set of ten and ☐ ones.

A set of ten and ☐ ones.

A set of ten and ☐ ones.

22 Ring the word that matches the numeral.

8	one three (eight) ten
10	three ten two nine
6	seven three zero six
4	four eight five ten
0	two zero nine one
2	ten three two six
9	nine one four eight
1	two six five one
7	zero seven three six
5	one five nine two

Ring the word that matches the numeral.   23

10	ten zero four two
7	eight nine six seven
2	five four two ten
8	six seven nine eight
5	five eight four zero
3	two three seven nine
0	ten five zero one
9	four zero seven nine
4	five four eight six
1	seven ten one three

24   Ring the numeral that matches the word.

| two | 8 | 6 | 10 | (2) | 7 |

| nine | 1 | 0 | 9 | 4 | 2 |

| zero | 8 | 6 | 3 | 5 | 0 |

| ten | 3 | 10 | 0 | 9 | 4 |

| six | 6 | 8 | 9 | 1 | 3 |

| one | 10 | 0 | 4 | 7 | 1 |

| eight | 5 | 3 | 8 | 1 | 6 |

| five | 1 | 5 | 10 | 8 | 0 |

| three | 8 | 2 | 9 | 3 | 10 |

| seven | 7 | 10 | 6 | 1 | 8 |

Ring the numeral that matches the word.

nine	7	10	5	9	8

one	8	1	9	6	3

four	0	9	4	2	6

ten	2	5	7	1	10

three	3	1	6	4	5.

seven	4	7	2	5	9

five	6	0	5	7	3

two	5	3	0	4	2

zero	10	2	9	0	7

six	7	9	4	2.	6

Ring a set of ten (10). Write the numerals.

A set of ten and [8] ones.

[ ] ten and [ ] ones.

A set of ten and [0] ones.

[ ] ten and [ ] ones.

A set of ten and [ ] ones.

[ ] ten and [ ] ones.

A set of ten and [ ] ones.

[ ] ten and [ ] ones.

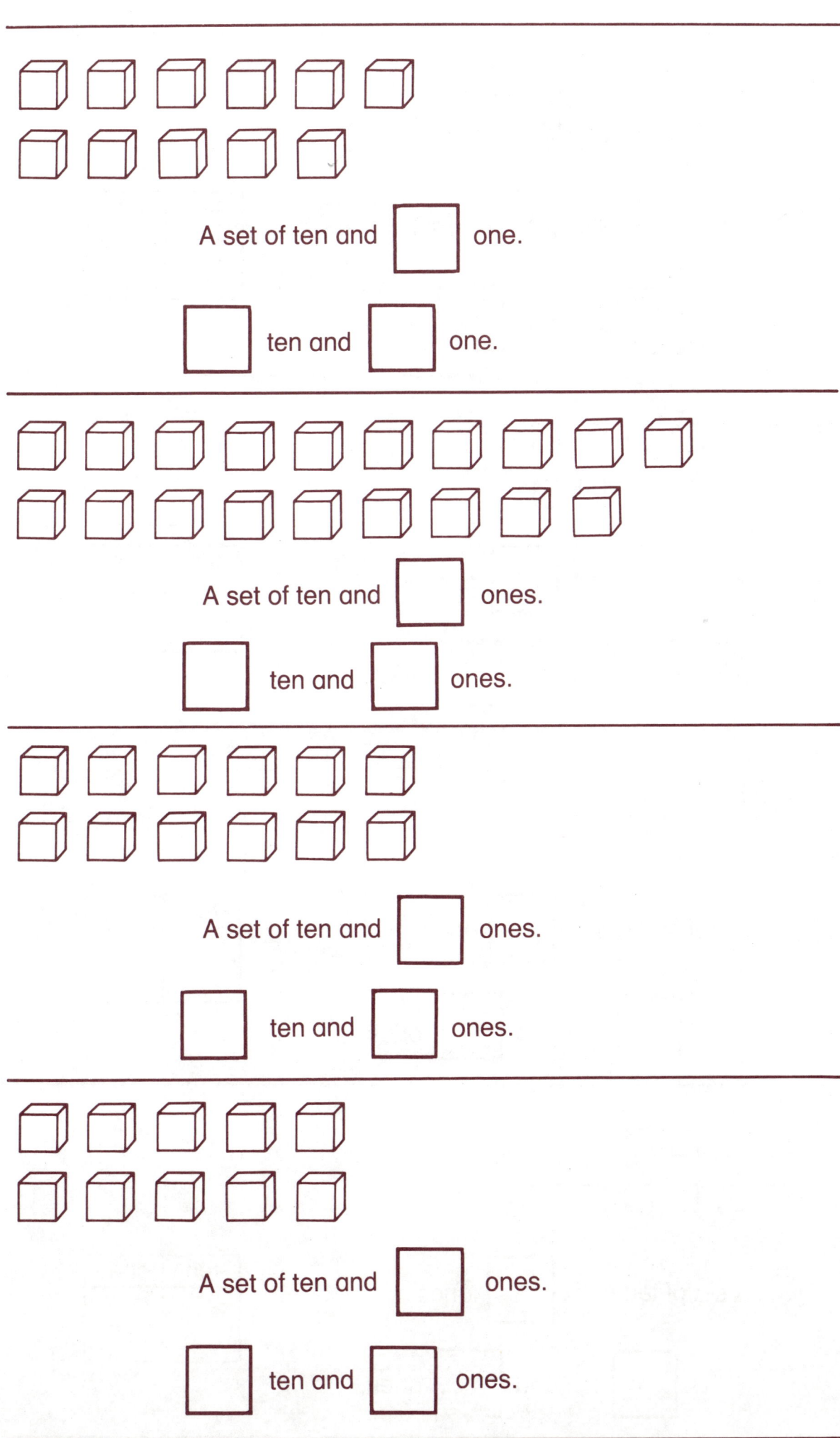

A set of ten and ☐ one.

☐ ten and ☐ one.

A set of ten and ☐ ones.

☐ ten and ☐ ones.

A set of ten and ☐ ones.

☐ ten and ☐ ones.

A set of ten and ☐ ones.

☐ ten and ☐ ones.

28   Ring a set of ten (10). Write the numerals.

▢ ▢ ▢ ▢ ▢ ▢ ▢ ▢ ▢ ▢
▢ ▢ ▢ ▢ ▢ ▢ ▢ ▢ ▢

A set of ten and ▢ ones.

▢ ten and ▢ ones.

ten	ones

▢ ▢ ▢ ▢ ▢ ▢ ▢
▢ ▢ ▢ ▢ ▢ ▢ ▢

A set of ten and ▢ ones.

▢ ten and ▢ ones.

ten	ones

▢ ▢ ▢ ▢ ▢
▢ ▢ ▢ ▢ ▢

A set of ten and ▢ ones.

▢ ten and ▢ ones.

ten	ones

▢ ▢ ▢ ▢ ▢ ▢ ▢ ▢ ▢
▢ ▢ ▢ ▢ ▢ ▢ ▢ ▢

A set of ten and ▢ ones.

▢ ten and ▢ ones.

ten	ones

□ □ □ □ □ □ □
□ □ □ □ □ □

A set of ten and ☐ ones.

☐ ten and ☐ ones.

ten	ones

□ □ □ □ □ □ □ □ □
□ □ □ □ □ □ □ □ □

A set of ten and ☐ ones.

☐ ten and ☐ ones.

ten	ones

□ □ □ □ □ □ □ □
□ □ □ □ □ □ □ □

A set of ten and ☐ ones.

☐ ten and ☐ ones.

ten	ones

□ □ □ □ □ □
□ □ □ □ □ □

A set of ten and ☐ ones.

☐ ten and ☐ ones.

ten	ones

Write the numerals.

tens	ones
1	4

tens	ones

tens	ones

tens	ones

tens	ones

tens	ones

tens	ones

tens	ones

Write the numerals.

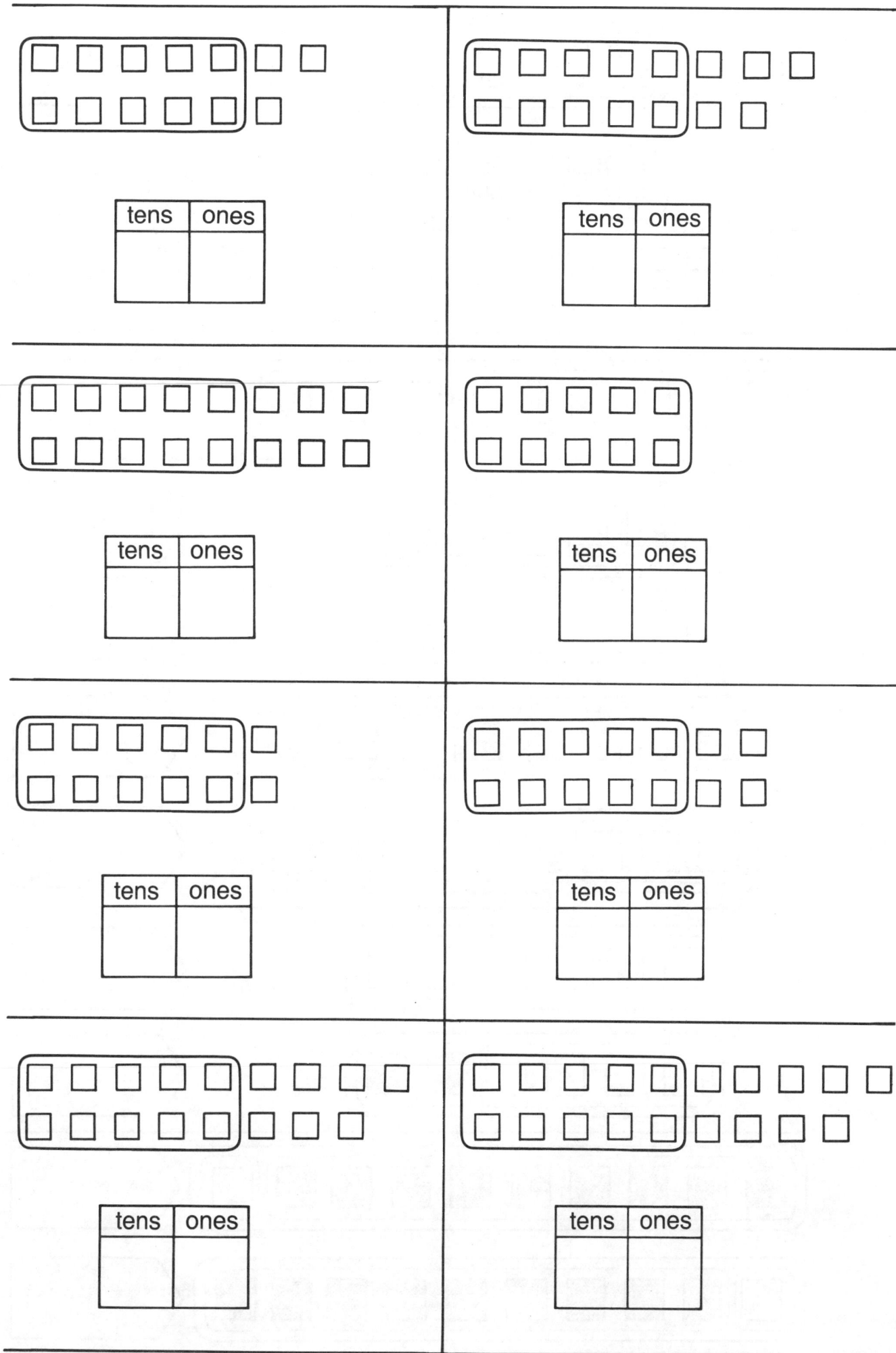

Write the number sentences.

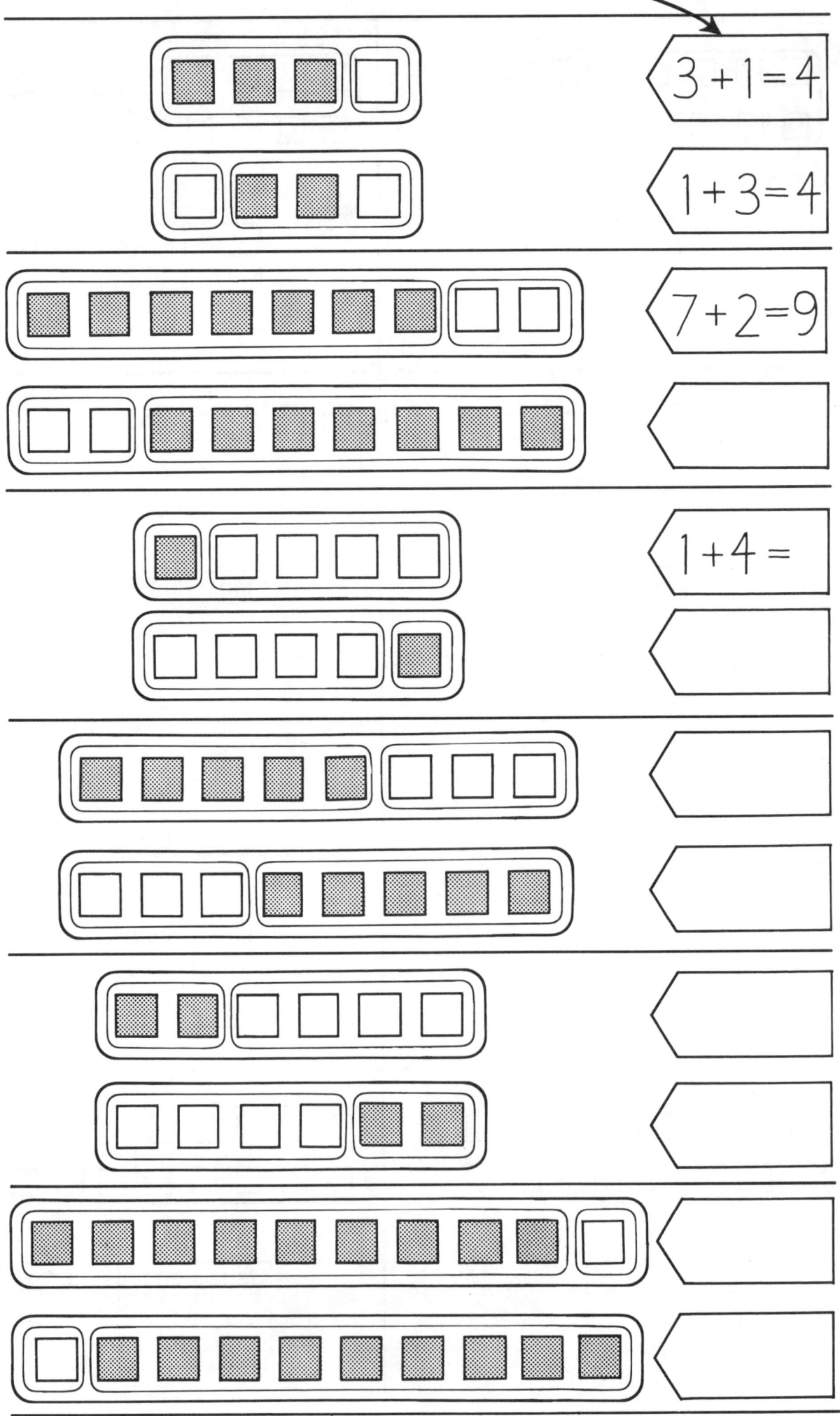

$3+1=4$

$1+3=4$

$7+2=9$

$1+4=$

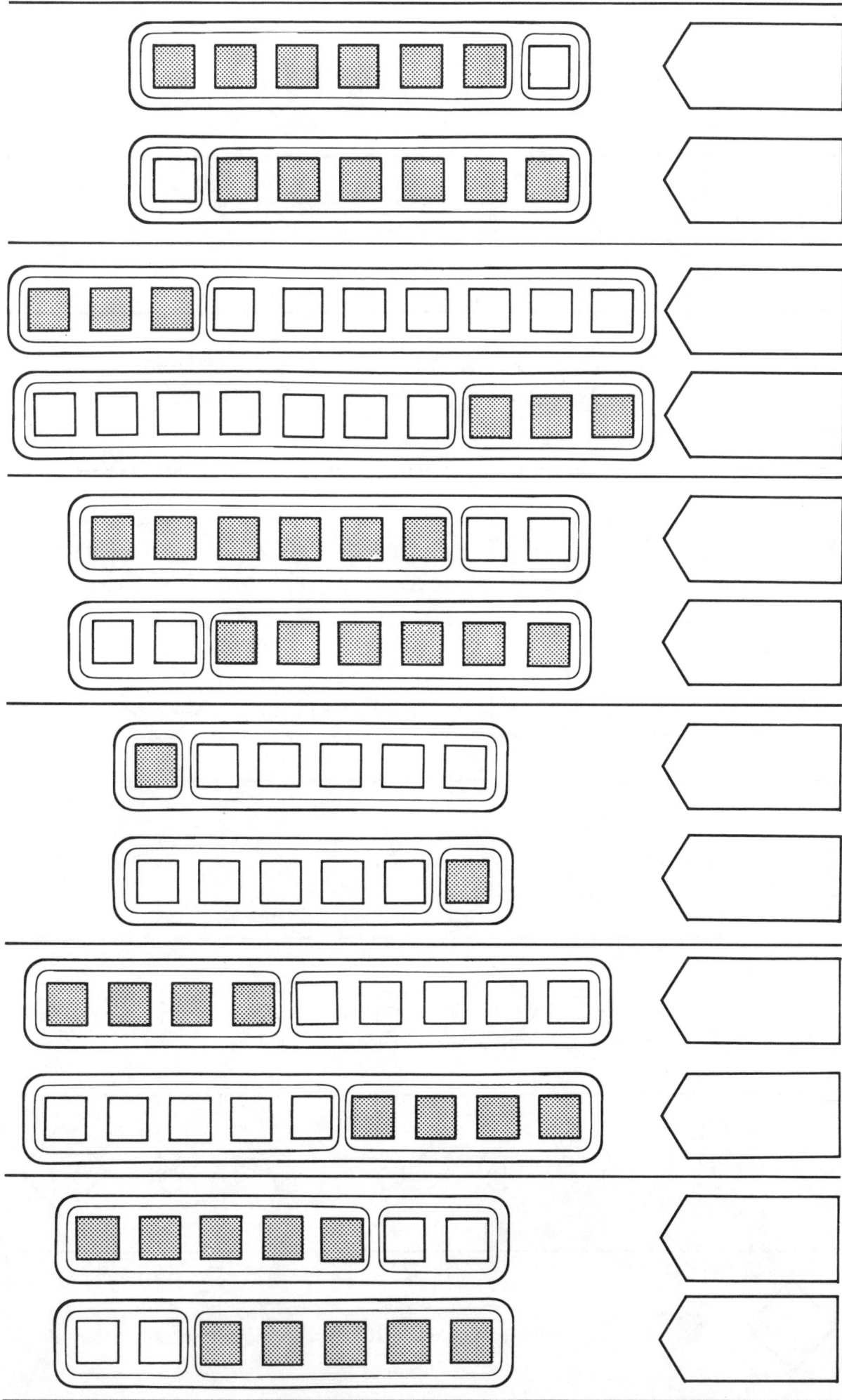

Write the missing numerals.

1 2 3 4 5 ◇6◇ 7 8

1 ◇ 3 4 5 6 7 8

1 2 3 ◇ 5 6 7 8

◇ 2 3 4 5 6 7 8

1 2 3 4 5 6 ◇ 8

◇ 2 ◇ 4 5 6 7 8

2 3 4 ◇ 6 7 8 9

2 3 4 5 6 7 8 ◇

◇ 3 4 5 6 7 8 9

2   3   4   5   6   7      9

2      4   5   6   7   8   9

2   3      5         8   9

3   4      6   7   8   9   10

3   4   5   6   7      9   10

3      5   6   7   8   9   10

   4   5   6   7   8   9

3   4         7   8      10

   4   5   6         9

Write the numerals.

There were 4 birds in the garden.

3 birds flew away.

There was then ⬡ bird left.

4 minus 3 equals ⬡

$4 - 3 = $ ⬡

There were 6 cakes on the table.

We ate 2 of them.

There were then ⬡ cakes left.

6 minus 2 equals ⬡

$6 - 2 = $ ⬡

James had 3 crayons.

He lost 1 of them.

He then had ⬡ crayons left.

3 minus 1 equals ⬡

$3 - 1 = $ ⬡

Write the numerals.

Sandra has 5 sweets.

She gives 2 to Tom.

She then has ⬡ sweets left.

5 minus 2 equals ⬡

$$5 - 2 = ⬡$$

There were 6 trees in the field.

The farmer cut down 3 of them.

There are now ⬡ trees left standing.

6 minus 3 equals ⬡

$$6 - 3 = ⬡$$

The baker had 4 loaves.

He sold 1 of them.

He then had ⬡ loaves left.

4 minus 1 equals ⬡

$$4 - 1 = ⬡$$

Write the numerals.

Ruth has 7 pencils.

Andrew has 5 pencils.

Ruth has 2 more pencils than Andrew.

7 minus 5 equals

$$7 - 5 = \bigcirc$$

Mark has 10 pence.

Debra has 6 pence.

Mark has 4 more pence than Debra.

10 minus 6 equals

$$10 - 6 = \bigcirc$$

Lucy has 8 marbles.

Adrian has 3 marbles.

Lucy has 5 more marbles than Adrian.

8 minus 3 equals

$$8 - 3 = \bigcirc$$

Write the numerals.

John has 9 sweets.

Sharon has 8 sweets.

John has 1 more sweet than Sharon.

9 minus 8 equals $\bigcirc$

$9 - 8 = \bigcirc$

Diane has made 6 cakes.

Tim has made 4 cakes.

Diane has made 2 more cakes than Tim.

6 minus 4 equals $\bigcirc$

$6 - 4 = \bigcirc$

Alison had 9 birthday cards.

David had 5 birthday cards.

Alison had 4 more birthday cards than David.

9 minus 5 equals $\bigcirc$

$9 - 5 = \bigcirc$

---

40   ## Write the numerals.

Mark has 8 pencils. Jane has 6 pencils.

What must be added to 6 to make 8?

$$8 - 6 = \boxed{}$$

So $\boxed{}$ must be added to 6 to make 8.

Ruth has 7 sweets. David has 4 sweets.

What must be added to make 7?

$$7 - 4 = \boxed{}$$

So $\boxed{}$ must be added to 4 to make 7.

Kevin has 6 toys. Julie has 1 toy.

What must be added to 1 to make 6?

$$6 - 1 = \boxed{}$$

So $\boxed{}$ must be added to 1 to make 6.

Joanne has 4 biscuits. Adrian has 3 biscuits.

What must be added to 3 to make 4?

$$4 - 3 = \boxed{}$$

So $\boxed{}$ must be added to 3 to make 4.

Write the numerals.

Nigel has 8 crayons. Hannah has 4 crayons.

What must be added to 4 to make 8?

$$8 - 4 = \boxed{}$$

So $\boxed{}$ must be added to 4 to make 8.

Tracy has 9 books. Steven has 3 books.

What must be added to 3 to make 9?

$$9 - 3 = \boxed{}$$

So $\boxed{}$ must be added to 3 to make 9.

James has 10 marbles. Sandra has 2 marbles.

What must be added to 2 to make 10?

$$10 - 2 = \boxed{}$$

So $\boxed{}$ must be added to 2 to make 10.

Lucy has 9 nuts. Martin has 2 nuts.

What must be added to 2 to make 9?

$$9 - 2 = \boxed{}$$

So $\boxed{}$ must be added to 2 to make 9.

Write the numerals to complete the number sentences.

$8 - 5 = \bigcirc$

$6 - 2 = \bigcirc$

$10 - 3 = \bigcirc$

$4 - 2 = \bigcirc$

$7 - 1 = \bigcirc$

$9 - 4 = \bigcirc$

Write the numerals to complete the number sentences.

$5 - 4 = \bigcirc$

$10 - 1 = \bigcirc$

$7 - 3 = \bigcirc$

$9 - 1 = \bigcirc$

$5 - 3 = \bigcirc$

$8 - 2 = \bigcirc$

44   Write the numerals to complete the number sentences.

$10 - 4 = \hexagon$ $6 + \triangle = 10$

$7 - 2 = \hexagon$ $5 + \triangle = 7$

$5 - 1 = \hexagon$ $4 + \triangle = 5$

$3 - 2 = \hexagon$ $1 + \triangle = 3$

$9 - 6 = \hexagon$ $3 + \triangle = 9$

$8 - 3 = \hexagon$ $5 + \triangle = 8$

Write the numerals to complete the number sentences.

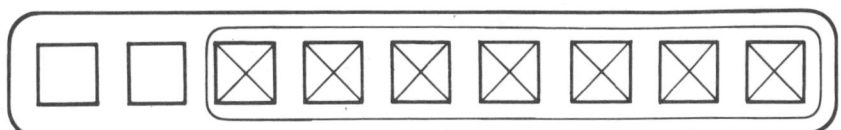

$9 - 7 = \bigcirc$ $\qquad$ $2 + \triangle = 9$

$5 - 2 = \bigcirc$ $\qquad$ $3 + \triangle = 5$

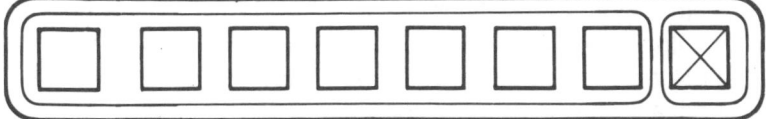

$8 - 1 = \bigcirc$ $\qquad$ $7 + \triangle = 8$

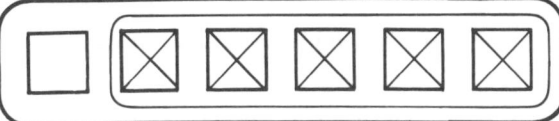

$6 - 5 = \bigcirc$ $\qquad$ $1 + \triangle = 6$

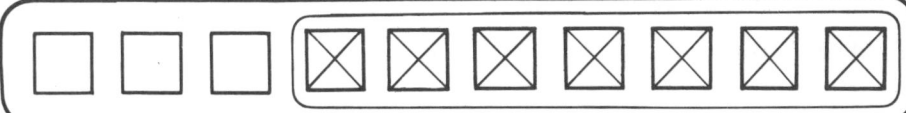

$10 - 7 = \bigcirc$ $\qquad$ $3 + \triangle = 10$

$7 - 6 = \bigcirc$ $\qquad$ $1 + \triangle = 7$

Oxford Introductory Maths Workbooks is a series designed to provide a gradual introduction to basic mathematical ideas, principles and language. Beginning at the pre-counting stage, the material focuses on the arrangement of familiar objects to lay the foundations of a real understanding of notation and what numbers mean.

This knowledge and awareness is then built on in order to bring out certain fundamental features of our number system – its additive nature, base, place value, conservation and reversibility and an understanding of the meaning and importance of zero. Throughout the course the material is presented in such a way as to encourage the pupils to discover these important elements for themselves by extracting the essential mathematical ideas from the activities in which they are embedded.

Since words are an integral and necessary part of mathematics these are introduced right away, though in strictly limited numbers at first and with adequate repetition. (Books 1 and 2 together employ a total vocabulary of only 61 words.) Mathematically correct terms are used in preference to apparently simpler but less precise words which may have to be 'unlearned' and discarded later on.

Special features of this scheme are:

☆ A systematic build-up of concepts

☆ The combination of discovery methods with the need for practice of basic skills

☆ Gradual progression and transition from one item to the next

☆ Early work consists of colouring and drawing and single-word responses

☆ Provision of ample reinforcement and revision

☆ Workbook format and layout is used to guide pupils towards accuracy and orderly habits of setting out their work

☆ Use of a limited, controlled vocabulary

☆ Each double-page opening is one unit of work

☆ Generous use of illustrations, diagrams and charts

☆ Clear examples at every stage

book 1 ISBN 0 19 918145 4

book 2 ISBN 0 19 918146 2

book 3 ISBN 0 19 918147 0

book 4 ISBN 0 19 918148 9

book 5 ISBN 0 19 918153 5

book 6 ISBN 0 19 918154 3

book 7 ISBN 0 19 918155 1

book 8 ISBN 0 19 918156 X

Oxford University Press
© Oliver Gregory
ISBN 0 19 918153 5

ISBN 0-19-918153-5

9 780199 181537